In the Driver's Seat

In the Driver's Seat

Ron & Nancy Goor

Thomas Y. Crowell   New York

ACKNOWLEDGMENTS

The authors wish to thank the following people and organizations for their help in making this book possible: *Blimp*, Goodyear Tire and Rubber Company; *Concorde Jet*, Frank Pirnat and Ewen Fraser, British Airways, and Roger Poore, John Litzenberger, Hugh Gudger, Henri Cloutier, Floyd Coddington, and Robert Schmitz, Federal Aviation Administration; *Fireboat*, D.C. Fire Department, Washington, D.C.; *Helicopter*, Metropolitan Police Department, Washington, D.C.; *M60 Tank*, Jim Coles, U.S. Army; *Race Car*, Polly Hammond, and Robert Berman, and the Sports Car Club of America; *Train*, Amtrak and Paul Cooley, Washington Terminal Company; *18-Wheel Truck*, William Shotwell, Jr., Ryder Truck Rental, Inc. and Sam Branum, C. R. Daniels Corporation.

Copyright © 1982 by Ron and Nancy Goor
All rights reserved. Printed in the United States of America. No part of this book may be used or reproduced in any manner whatsoever without written permission except in the case of brief quotations embodied in critical articles and reviews. For information address Thomas Y. Crowell Junior Books, 10 East 53rd Street, New York, N.Y. 10022. Published simultaneously in Canada by Fitzhenry & Whiteside Limited, Toronto.
Designed by Harriett Barton

Library of Congress Cataloging in Publication Data
Goor, Ron. In the driver's seat.
Summary: Photographs and text present a driver's-eye-view of operating an eighteen-wheeler, combine, train, blimp, front-end loader, Concorde jet, crane, race car, and tank.
1. Motor vehicle driving—Juvenile literature.
2. Airplanes—Piloting—Juvenile literature. [1. Motor vehicle driving. 2. Airplanes—Piloting] I. Goor, Nancy.
II. Title.
TL147.G63 1982   629.04′6   81-43885
ISBN 0-690-04176-4   AACR2
ISBN 0-690-04177-2 (lib. bdg.)
10 9 8 7 6 5 4 3 2 1
First Edition

To our parents,
Jeanette and Charles,
Helen and Martin—
with love

Contents

Front-End Loader	1
Combine	9
Blimp	17
M60 Tank	25
Race Car	33
Concorde Jet	41
18-Wheel Truck	53
Crane	63
Train	69
Other Driver's Seats	79

Front-End Loader

Like a gigantic metal monster, the front-end loader takes a bite out of the side of a wrecked building. The machine's huge jaws tear at chunks of bricks, cement, and steel rods. Clouds of dust fly up from the rubble. The loader quickly backs away from the wall. The machine moves so fast that it seems to slide backward down a hill of dirt. Rumble, rumble. There is a constant roar as the loader does its work.

You are the driver of a front-end loader. You sit in an open cab behind the powerful jaws that are called the bucket. It takes skill to drive the loader. You must move a 48,400 pound machine forward, backward, and to the left or right, and at the same time you must raise, lower, open, or close its bucket.

There are three pedals on the floor of your cab. You use the left and right pedals to steer. The middle pedal is an emergency brake. The large lever to the left of your seat makes the front-end loader go forward or backward. The smaller lever regulates the speed. This lever is like the gas pedal in an automobile. The levers on your right operate the bucket. The large lever moves the bucket up and down. The small lever opens and closes the claws of the bucket. With these controls you can lift 35,000 pounds.

You lower the bucket and scoop up the rubble. Like a monster with a mouth full of metal and concrete, the front-end loader stretches its neck high, opens its jaws, and drops its load into a waiting trailer dump truck. Boom. Crash. Metal meets metal. The loader's bucket hits the trailer truck and the truck shakes violently.

At the end of the day you climb down from your

seat in the cab. You are covered from head to toe with a layer of dust. The dust is so thick someone could write on your forehead. As you leave the work area you walk by piles of broken cinder block and chunks of cement. Tomorrow you and your metal monster will finish clearing the site. It takes you only a few days to clear away a building that took months to construct.

Combine

It is a clear, bright, hot summer day. The sun beats down on the farmhouse, on the fields of wheat, and on the bright red combine. You use the combine to harvest wheat, corn, and soybeans. Today you will harvest wheat.

You climb up the steps into the glass-enclosed cab and slide into the driver's seat. You put the key in the ignition and press the starter button. The combine begins to vibrate and roar. The two big, odd-shaped handles at the very front of the control panel are gearshifts. You put the combine into gear. Behind the gearshifts are three levers. The middle one is the hydrostatic lever. It makes the combine go forward

or backward, fast or slow. You push the hydrostatic lever forward. The combine lurches ahead. You push the hydrostatic lever farther forward. The combine goes faster.

The combine is a big machine. It is very wide. You turn the steering wheel to move this 12-ton giant carefully through the farm gate to the field. A thick blanket of wheat covers the earth. You press the lever at the left of the hydrostatic lever to lower the head at the front of the combine. The head contains, on top, a reel and, at the bottom, a 15-foot-long knife with a blade like a saw. Above the knife is an auger. The reel turns around and around and throws the wheat against the knife. The knife slides from side to side, cutting down stalks of wheat. The auger, which is shaped like a drill bit, revolves and moves the wheat out of the head onto a conveyor

belt. From there the wheat moves into the body of the combine. Inside, the grain is separated from the chaff and straw.

You move the combine forward. The motor roars. The seat vibrates. From your seat in the cab you watch the wheat being gobbled up by your machine. The chaff and straw pour out behind the combine, leaving long, flat rows where wheat once stood. The

grain goes into a storage tank inside the combine. The combine rattles, wheezes, and whirs along, mowing down row after row of wheat.

Soon the grain storage tank is full. You press a button to raise the head high above the ground, and drive to a truck parked at the edge of the field. You push the lever at the left of your seat to move the unloading auger over the open trailer. The

long, metal arm swings into position. Grain pours out of the auger in a steady stream. It pours and pours, filling part of the truck.

When the grain storage tank is empty you back up the combine, turn it around, and roar back to the field. It is only 10:30 A.M. and you have 300 acres of wheat to harvest before the sun sets.

Blimp

A strange, football-shaped balloon sails slowly across the sky. From the ground it looks small, but it is enormous. It is 192 feet long, as long as a Boeing 747 Jumbo Jet. It is 59 feet high, as tall as a six story building. It is 50 feet wide, as wide as a basketball court. What is it? Is it a plane? Is it a balloon? No, it is a lighter-than-air aircraft called a blimp.

You are the pilot of the blimp. You sit in the left-hand seat at the front of the car, which is a small cabin below the balloon. Today you are taking passengers on a short flight around Dulles Airport in Washington, D.C. The car is small. There is only enough room for you and six passengers.

You start the engines. Because a blimp is filled with helium gas and is lighter than air, it does not need powerful engines to lift it off the ground and keep it afloat. The two 210-horsepower engines mounted behind the car push the blimp lazily through the air at 35 miles per hour.

The elevator wheel is at the right of your arm rest. When you rotate it backward, the elevators or flaps on the horizontal fins at the rear of the blimp go up. The tail goes down, and the nose lifts up. You push the white-handled throttles at the left of your seat forward and the engines speed up. Outside, members of the ground crew grab onto the sides of the car and shove the blimp into the air. Up, up it goes. The blimp rises at such a steep angle you are thrust back against your seat.

You check the gauges and dials on the instrument

panels above and below your window. You keep a close watch on the gauges that show the air and helium pressure inside the balloon. It is this pressure that gives the balloon its shape. Hidden inside the balloon are two big airbags called ballonets. As the blimp rises, lower air pressure outside the blimp causes the helium inside the balloon to expand. To make room for the expanded helium you must let air out of the ballonets. You control the amount of

air in the ballonets with the toggle switches on the panel above the windshield.

When you reach 1,000 feet you let go of the elevator wheel and it rotates forward by itself until the blimp levels off. You push the left rudder pedal down with your foot and the flaps on the vertical fins at the rear of the blimp move to the left. The blimp slowly floats in a big circle. You feel as if you are sitting in a boat that is gently rocking in place.

Soon the Dulles tower comes into view again. Time to descend. You push the elevator wheel forward. The blimp floats slowly to the ground. As you descend, the helium inside the balloon contracts and takes up less space. To prevent the balloon from collapsing, you must add air to the ballonets. The ballonets expand and fill the space that the helium once occupied.

When you are within 60 feet of the ground six members of the ground crew run toward the blimp. They catch the two ropes hanging from its nose. They pull the noselines with all their might until the blimp's nose faces into the wind. If the blimp's gigantic side were to face the wind it would easily blow the blimp back up into the sky.

Thud! The small tire beneath the car hits the ground. A member of the ground crew climbs to the

top of the mooring mast and attaches the blimp's nose to the tip of the mast. Sandbags are hung from a bar at the bottom of the car. Now the blimp will not float away. You turn off the engines. You and your passengers climb down from the car. In a few hours you will take this super balloon into the air again for an aerial tour of Washington, D.C. at night.

M60 Tank

The U.S. Army's M60 battle tank is tremendous. It weighs 105,000 pounds. Its metal treads allow this monster to ford deep rivers, cross rough ground, and climb easily over steep hills. Even though the tank is large, from a distance it is difficult to see. It is painted dark green with patches of brown so it will blend in with its surroundings.

The M60 is a war machine. It has a 105mm gun mounted on a rotating turret above its hull or body. A 50-caliber machine gun is also mounted on the turret. A smaller machine gun sits beside the cannon barrel. The turret can turn in a complete circle so the guns can shoot in any direction.

You are the driver of the M60 tank. You drive the tank to the battleground and maneuver it during the fighting. You do not shoot or load the guns. The commander and other members of the crew—the gunner and gunloader—do these jobs.

You sit in a small compartment in the hull of the tank. To enter the driver's seat from the turret, hold onto the shelf above the opening and swing your body down, feet first. Drop gently onto your cushioned seat. The driver's seat is almost resting on the floor. You are enclosed by thick walls of armor-plated steel. Adjust a lever on your left and the chair jerks up a few inches. Now you can see through the only windows in the hull—three narrow periscope windows in a row just below the ceiling. On the outside of the tank the windows are just below the turret.

To start the tank you flick a master switch on the

control panel at your right. When you press the starter button the diesel engine begins to hum. You shift the lever on your right to put the tank into high, low, or reverse gear. The pedals on the floor are similar to those in a car. The left pedal is the brake. The accelerator pedal on the right controls the tank's speed, 10-12 miles per hour on rough ground and 25-35 miles per hour on smooth road. You use the

T-bar directly in front of you to steer the M60. Push the bar to the right and the tank turns left. Push the bar to the left and the tank turns right. To make the tank pivot, push the T-bar all the way to one side while you step on the accelerator.

The control panel at your right also has switches to regulate the tank's lights. When it is dark outside, you often turn on special night-vision lights called infra-red lights, which the enemy can't see. The light can only be seen with a special viewer. When you look through the viewer, the road ahead looks as bright as day.

"Target...two o'clock...range." You hear your commander talking to the gunner on the intercom. The gunner swings the gunsight and answers, "Target confirmed...range 1,000." The gunloader quickly loads a 54-pound shell into the 105mm gun. The

gunner fires. Boom! A flash of fire! A burst of smoke! Inside the tank you feel a sudden jerk. Three seconds later another shell is fired. Boom! The target is hit.

Today you and your fellow crew members are practicing on a firing range. The commander sits in the open turret hatch so he can see if the target has been hit. In battle the crew stays inside and all hatches are kept closed.

A tank is both a safe and a dangerous place to be. It is safe because it is so big and heavy. Bullets bounce off its armor plates. It is dangerous because if the tank is hit by an enemy shell, and then explodes or catches fire, the crew is trapped inside. In battle the tank crew must work closely together. Their survival depends on it.

Race Car

You are a race car driver. You drive an open wheel formula Ford racing car. Your car is light so it will go fast. It is wide and low so it will hug the road. It will not tip over easily as it turns corners at high speeds. It will meet with little air resistance as it roars along the roadway.

Five minutes before race time you ease yourself into the specially molded seat on the floor of your race car. You stretch your legs straight out in front of you, lean back, and fasten your seat belt. Your car is designed so you can feel the car's movements in your back and legs. You can feel where the raceway is smooth or bumpy, flat or banked. You feel as if you are part of the race car.

One minute to race time. You place your right hand on the gearshift at the right of your steering wheel and your right foot on the gas pedal. You rev the engine, shift into gear, and enter the raceway.

The green flag comes down and the race is on! The noise is deafening as the 26 cars roar off.

You streak down the straightaway at 135 miles per hour. Here comes the first curve. Slow down to

SUMMIT POINT RACEWAY

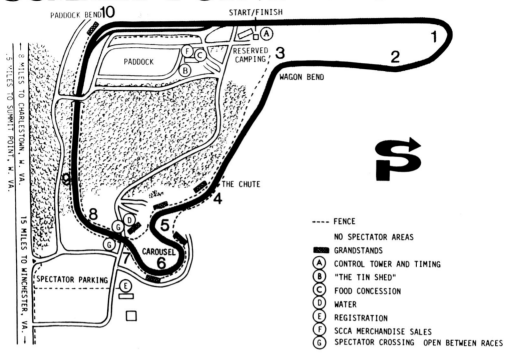

make this sharp right turn. A quick glance at the tachometer in the center of your dashboard shows you how fast the motor is turning. When the dial reaches 6,000 revolutions per minute, you shift down from fourth to third gear. Screech! You slow down to 70 miles per hour. As you round the curve you accelerate. When the tachometer again reaches 6,000 revolutions per minute you shift back to fourth

gear. You shift quickly so you do not lose speed. You take a quick look at the oil pressure dial on the left side of the dashboard. The oil pressure is OK. No leaks. A glance at the right-hand dial shows you the water temperature is not too high. You watch the communicators at the flag stands. The communicators use different colored flags to signal the racers.

Here comes curve number five, the trickiest curve on the race course. A communicator is waving a yellow-and-red-striped flag. Watch out! There is oil on the road. Car 53 is spinning around on the oil. Because race cars sometimes crash and catch on fire you are wearing special fireproof clothes—a one-piece fireproof jump suit, a fireproof mask under your helmet, and fireproof gloves. Even your underwear and socks are fireproof. The driver of Car 53 is not hurt. You race on.

You slow down as you approach a sharp and dangerous curve, curve number one. Car 18 is ten feet ahead of you. He screeches around the curve. You screech around the curve, accelerate, and pass him. You see your pit crew captain holding a number 4 sign by the side of the raceway. You have four more laps to go. Round and round the race course you go. You are coming into the final lap. You are in third place. You speed up for the last lap. You get close to Car 54, but you cannot pass him. Car 44 zooms across the finish line. The checkered flag swings down. The winner! Car 54 comes in second. You come in third. You drive back to the pits. You wish you were the winner, carrying the checkered flag on the victory lap around the race track. Maybe next time.

Concorde Jet

You are the pilot of a Concorde supersonic jet, the fastest passenger plane in the world. Today you are flying 100 passengers 3,423 miles. You are travelling from Dulles International Airport in Washington, D.C., to Heathrow International Airport in England.

You sit in the seat on the left and your co-pilot sits in the seat on the right. There are identical control panels in front of both seats. On the panel directly in front of you is the attitude dial. The position of the artificial horizon tells you if the plane is climbing or descending, or turning to the left or right. Below the attitude dial is the dial that tells you the direction in which the plane is

flying. To the left of the attitude dial is the mach meter, which indicates the plane's speed. Mach 1 is the speed of sound, about 700 miles per hour at sea level. Mach 2 is twice the speed of sound or about 1,400 miles per hour.

The dials on the center panel tell you the engine speeds and temperatures. They show you the flow of fuel to the four engines. Below the front windows are the autopilot switches, which can automatically control the plane from takeoff until landing. The console between you and the co-pilot contains all the radios for communication and the navigation instruments. The four large handles in the middle of the console are throttles to regulate the speed of the four engines.

After the passengers and baggage have been loaded onto the plane the doors are closed. At the direction of a mechanic on the ground you and your crew start

the four engines, one at a time, using overhead switches. You push the visor lever to tilt the hinged nose of the plane down five degrees. This allows you to see the runway as you taxi and take off. You radio the control tower, "Speedbird 188 is ready for take off."

"Speedbird 188, taxi into position and hold," replies the air-traffic controller. You must wait for incoming planes to land and planes ahead of you to take off.

Soon the controller radios to you, "Speedbird 188 cleared for take off."

You push the throttles forward. The supersonic jet begins to race, faster and faster, down the 11,500 foot runway. The co-pilot calls out when you have reached 180 miles per hour. After this speed your plane is going too fast to stop on the remaining length of runway. You must take off. When the plane reaches 220

45

miles per hour, you pull the handlebars of the control column between your legs toward you. The plane noses up into the air. As you climb into the clouds you retract the landing gear.

Departure Control gives you instructions for leaving the Washington area. "Speedbird 188, turn right direct Baltimore, climb and maintain flight-level two three zero."

You turn the control column to the right to bank the plane. You head toward the Baltimore radio signal. Baltimore is the name of one of the VORTAC radio navigation signals that give pilots exact information on their position as they fly over land. You are climbing toward 60,000 feet and heading east. This is twice as high as regular passenger jets fly. At 60,000 feet you are above the storms and bumpy weather. The flight is very smooth.

You are now over the Atlantic Ocean flying at 630 miles per hour. You push the throttles forward more, edging the plane toward mach 1. In order to pass through the sound barrier you must use the afterburners to provide extra power. Neither you nor your passengers feel the plane pass through the sound barrier, but if you were outside the plane you would hear a tremendous boom. Supersonic jets always pass through the sound barrier over the ocean so the shock wave is absorbed by the water and will not disturb people on land.

As you climb through 50,000 feet the plane is flying at mach 2 (1,400 miles per hour), or twice the speed of sound. The maximum speed of the Concorde is limited by the temperature to which the outside skin of the plane heats up. At a speed above mach 2 the skin of the plane becomes too hot for safe operation. You

and your crew carefully watch the temperature dials. Your engineer also adjusts the air conditioning to keep the temperature inside the airplane comfortably cool.

The fuel is stored in 13 tanks in the wings and body of the plane. Your engineer moves fuel from tank to tank during the flight to keep the plane balanced.

You switch the plane onto autopilot. For the next three hours the autopilot will automatically navigate the plane, according to the route you entered into the navigational computer.

When you near Heathrow Airport you tune your radio to the Airport Automatic Terminal Information Service. You learn the weather at the airport and some general information about landing.

"Heathrow International Airport Information Alpha measured ceiling two thousand overcast, visibility four…temperature six five…wind one five zero degrees at two five…altimeter two niner niner eight …ILS runway two eight L approach in use."

The controller at the airport advises, "Speedbird 188 descend and maintain 3,000 feet…report leaving 7,000 feet."

You ease forward on the control column. The plane noses downward and begins to descend. You tilt the nose of the plane down so you can see the runway. Following the radio beam of the instrument landing system you softly set the supersonic jet down on runway 28L. In less than four hours, half the time it takes a subsonic jet, you have flown 100 people 3,423 miles across the Atlantic Ocean, from Washington, D.C., to London.

18-Wheel Truck

The blue and white 18-wheel truck is loaded and ready to go. The truck is very big. It is 45 feet long, 13½ feet tall, and weighs 37,000 pounds unloaded. Today it is filled with cartons of canvas and weighs 77,000 pounds.

You are the driver of this 18-wheel truck. Today you will drive 500 miles from Ellicott City, Maryland, to Rutledge, Tennessee. Before you set out you check the air brake hoses and the electrical wires between the trailer and the cab. You check the gas tanks—each of which holds 135 gallons of diesel fuel, enough to drive 1,350 miles. You hit each tire with a metal rod and listen for the sound that tells you the air pressure is right.

You climb into the wide cab and ease into the driver's seat. You have a special seat that slides back and forth on a track and moves up and down on a cushion of air. Without this special seat you would feel every bump in the road as the heavy trailer bangs into the back of the cab. You turn the key and push the engine button to start.

Because your truck is so heavy you need powerful air brakes to make it stop. You check the button in the center of the dashboard. When the button lights up the air pressure in your brakes is too low. The light goes off when the air pressure reaches 70 pounds per square inch. You check the air pressure gauges at the top right of the dashboard. Hiss…hiss…the pressure is building. Hisssss! The pressure builds up to 120 pounds per square inch. Now you can release the brakes and move the truck.

(Smokey Bear) are at the ten-mile marker on the southbound side of the road using radar (shooting) to measure the speed of cars (four-wheelers) from behind (in the back). You always drive at the speed limit, but you answer, "Thank you. Have a safe one."

You start up a steep hill. You pass an 18-wheeler moving slowly in the right lane. Your truck is so long it is hard for you to judge when you have completely passed the other truck. You look in the large mirror

outside your right window. The other truck driver flashes his headlights. Now you know it is safe to pull back into the right lane. You blink the body lights on your truck to thank him.

Your stomach growls. Time for dinner. You pull into a truck stop, carefully easing your truck into an empty parking slot. Your truck is 45 feet long, but you drive it as if it were a sports car. You have no trouble backing up, turning corners, or parking.

After dinner you climb back into the cab. You drive all night. You arrive in Rutledge at sunrise, riding down quiet streets, past churches and schools and houses with white picket fences. You drive into the factory parking lot at the edge of town. You jump down from the cab and open the rear door of the truck. You climb

in the cab again and carefully back up the truck to the loading dock. Then you shut off the motor and crawl into the bed behind your seat. While you sleep factory workers unload the truck. You must sleep soundly because in several hours you will climb back into the driver's seat and drive 500 miles home.

Crane

You are a crane operator. You lift yourself onto the crane platform. You climb into the cab of the crane, which is behind the 40-foot-long boom. The cab is very small. It only has room for your seat and the levers that control the crane. You slide onto the seat sideways and pull the door shut. It is cozy and warm in the heated cab. The workmen outside are not so lucky. Today it is windy and cold.

You turn the key to start the engine. The motor is powerful. It makes a loud purring noise. The whole cab vibrates.

On the dashboard there are four large levers. On the floor there are two pedals. With these controls you can wreck an entire building. You can also pick

up large pieces of the wrecked building and drop them into a waiting dump truck. The lever on the left swings the boom and cab to the left or right or in a complete circle. The lever on the right moves the boom up and down. The middle levers raise and lower the two hoists that hang from the boom. The two pedals are brakes for the hoists.

A two-ton steel ball is attached to one of the hoists today because the crane is being used to wreck a three-story building. To knock down a wall of the building you push the left hand lever to the left. The ball swings away from the wall. Then you push the same lever to the right. The crane jumps and shudders. The ball swings back and crashes into the wall. Large pieces of concrete and steel fly out and tumble to the ground. A cloud of dust appears, and when it settles you see a large hole in the wall. Sometimes

a whole wall collapses with one hit of the ball.

To knock down the floors and the roof you move the middle lever to the right. The hoist raises the ball high into the air. You put your foot down on the right brake pedal and the ball stops. You release the brake pedal and the ball drops, smashing the roof. Boom.

The crane jumps. The crane shakes. You feel as if the crane is alive.

The crane is a powerful machine. You must work carefully so you do not damage buildings near the one you are wrecking. You must be alert at all times, have quick reflexes, and have good aim.

Train

The mighty steel engine waits silently in the train yard. Its streamlined silver body gleams in the sunlight. In less than an hour it will pull eighteen cars filled with hundreds of passengers from Washington, D.C., to New York City.

You are the engineer of this E60 locomotive. You climb up into the cab of the engine. You sit in the seat on the right side of the cab. Your control panel is arranged at a slant so you can watch the gauges and operate the controls while you look out the front or side windows. Your locomotive is electric. You press a button to raise the elbow-shaped pantographs on top of the engine to the wire (called a catenary

wire) high above the track. Electricity flows from the wire through the pantographs to power the locomotive.

You always carry a set of handles with you. You insert one handle into the reverser slot directly above the row of five white buttons on the control panel. The reverser starts the engine and makes the engine go backward and forward. At the left of your seat, just below the telephone, is the automatic brake slot. You place another handle in this slot and turn on the brake. Hiss …hiss…the air pressure begins to build inside the brake lines. The automatic brake handle applies the brakes in the engine and in all the cars. You put the last handle into the independent brake slot, below the automatic brake slot. The independent brake handle turns the brakes on and off only in the engine.

Ding-ding-ding-ding, you drive the engine into the station. Eighteen shiny silver passenger cars wait on

the track, ready to be coupled to the big engine. Ten feet from the first car you bring the engine to a stop. Slowly, slowly, you move it toward the passenger car. Bang! The two steel giants couple.

Now a carman attaches a blue light beneath the cab window. The blue light means "Danger! Men at work between the cars. Do not move the train until I take the light away!" Workers connect the airbrake hoses

and electrical cables between the engine and the passenger cars. You apply the brakes and the carman checks every car to make sure the brakes go on.

Hiss…hiss…The 10 A.M. Congressional to New York City is ready for departure.

You are part of a big team. The conductor is the boss. He makes sure that the passengers are safely on

board and the train is ready to go. When it is, he signals the block operator, who sits in a control tower along the tracks. The block operator controls train traffic in the station area, setting switches and changing signals by remote control. Now he changes the electric signal to tell you that the track is clear. But, still you wait. The conductor must also give you the signal to go. After he signals, you move the reverser handle forward, release the automatic brake, and move the throttle slowly to number three. The train begins to move. Slowly, slowly you snake out of the station. You follow the signals and switches that the control tower has set for you.

You have a very important job. You must carry hundreds of passengers safely to their destinations. You must always be alert. A special system has been designed to help you stay alert. It is called the Deadman or Alertness Control. Throughout the trip you must touch the control panel in a different place every 15 seconds. If you fail to do this, a siren sounds in the cab, a light blinks in your eyes, and the brakes automatically stop the train.

Again, you check the signal box above and between the front windows. The lights that appear here tell you if the track ahead is clear or occupied by another train.

You see two small lights at a slant. This means restricting approach—go slowly—maximum speed 20 miles per hour.

The lights in the signal box change. Three small lights in a vertical line appear. All clear. You move the throttle to number eight. Full speed ahead to New York City!

Fireboat

Forklift

Seattle Monorail

Helicopter

San Francisco Cable Car

Ron Goor was born and raised in Washington, D.C. He was graduated from Swarthmore College and attended the University of Chicago and Harvard University, where he received a master's degree in public health and a Ph.D. in biochemistry.

Nancy Goor also grew up in Washington, D.C. A graduate of the University of Pennsylvania, she received her M.F.A. from Boston University.

The Goors are the authors of *Shadows: Here, There, and Everywhere,* an American Library Association Notable Children's Book of 1981.